INTRODUCTION

The desirability of a snake often is as dependent on its history as on its appearance. With time, some species develop a mystique that far outweighs their beauty or keepability. This is the case with the Gray-banded Kingsnake, *Lampropeltis mexicana*, a species from Texas and northern Mexico with a long history as a rare and unique species that remains poorly known in nature but has become relatively common in the reptile hobby. Though the species, with its various color phases and synonyms, was known from perhaps just two or three dozen specimens in 1960, only 35 years later it is possible to purchase captive-bred specimens of almost all patterns found in the species and in its close relative *Lampropeltis ruthveni*, Ruthven's Kingsnake. An amazing number of breeders and hobbyists keep these species, and many specialize in

The Gray-banded Kingsnakes, here represented by an unusual *Lampropeltis m. mexicana* with a reduced pattern showing some influence from *L. m. alterna*, are among the most colorful snakes in the hobby today.

PHOTO: M. PANZELLA.

collecting the numerous variations of colors and patterns known in nature and in the hobby.

This book is the first attempt to provide a comprehensive coverage of the Gray-banded and Ruthven's Kingsnakes. In addition to the usual chapters on care and breeding, I will explore the history of the species a bit, including some rather complicated taxonomy and abrupt changes of opinion by several specialists. There is surprisingly little literature on these species when you consider how important they have become in herpetoculture, and much of it is confusing to the uninitiated. There are basic questions to be answered as to the number of species and subspecies that could be recognized in these snakes and the relationships of various described forms that are treated as species by hobbyists and as simple synonyms by taxonomists. Even the forms that occur in Texas and are collected in rather large numbers each year are not fully understood. This book will not solve the problems about relationships of the various Gray-banded Kingsnakes, but I hope that it will give you a good introduction to these wonderful snakes and a better understanding of variation in the species.

WHERE THE GRAY-BAND AND RUTHVEN'S FIT

The kingsnakes, genus *Lampropeltis*, are an American specialty. The seven species commonly recognized range from the U.S.–Canadian border area

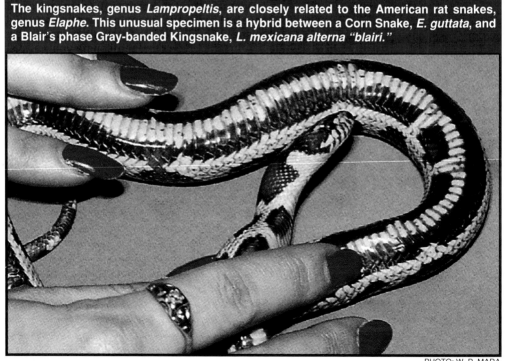

The kingsnakes, genus *Lampropeltis*, are closely related to the American rat snakes, genus *Elaphe*. This unusual specimen is a hybrid between a Corn Snake, *E. guttata*, and a Blair's phase Gray-banded Kingsnake, *L. mexicana alterna* "blairi."

PHOTO: W. P. MARA.

INTRODUCTION

The *getula* group of *Lampropeltis* contains two species that typically lack red in the color pattern and have short teeth at the rear of the upper jaw. The very popular California Kingsnake, *L. getula californiae*, here in the "banana split" phase, is typical of this group.

south over virtually all the United States and Central America into northwestern South America. These snakes are very closely related to both the American rat snakes, genus *Elaphe*, and the gopher snakes, genus *Pituophis*, and they probably share a common ancestor. Because species of these three genera can be successfully interbred in captivity (though hybrids are not known in nature), their chromosomes are virtually identical, and they show very few differences in cellular chemistry that can be detected by electrophoresis and similar chemical tests, it has been suggested that they belong in a distinct tribe of the family Colubridae, the Lampropeltini. In structure *Lampropeltis* is defined by a distinct lack of specializations, being just big, smooth colubrid snakes with normal head plates, 17 to 27 rows of scales across the back, two apical pits on most dorsal scales (most readily seen in shed skins), and a single (undivided) anal scale. The hemipenis is weakly bilobed, has a single sulcus spermaticus (sperm groove), and has only small spines about the middle. For a good overview of the genus *Lampropeltis*, see Ron Markel's book *Kingsnakes and Milk Snakes* (T.F.H. TS-125).

Once known as *Ophibolus*, a name no longer being used, the kingsnakes fall into two rather distinct groups based on coloration, hemipenis spines, and dentition. The typical kingsnakes, represented by *Lampropeltis*

INTRODUCTION

TS-125
Kingsnakes and Milk Snakes

getula and *L. calligaster*, typically lack red in the pattern (especially as bands over the back), have fairly large and regular spines over the outer third of the hemipenis but lack spines on the basal half or so of the hemipenis (which is distinctly bilobed), and have the maxillary teeth rather small and all of similar size. The milk snakes, which could be placed in the subgenus *Osceola*

Some Mole Snakes, *L. calligaster rhombomaculata*, may be quite reddish for their first year or so, then they turn brown. Notice the strong and distinctive pattern on the head in this juvenile, a feature always different from the head pattern of Gray-banded Kingsnakes.

Baird & Girard, tend to have color patterns containing red rings or saddles over the back, have small

This mahogany-tinted Common Kingsnake, *L. getula*, from southern Florida (subspecies *brooksi*) is about as red as this species gets.

PHOTO: K. H. SWITAK.

INTRODUCTION

The Milk Snake *L. triangulum arcifera* may be found with Ruthven's Kingsnake and is difficult to distinguish from it without scale counts. The whitish snout and tendency toward broken rings, as well as some black rings expanded over the center of the back, may help distinguish this snake from *L. ruthveni*, but you probably will have to count ventral scales to be sure: the Milk Snake has 192 to 217, Ruthven's 182 to 196.

spines in a broad band over the central third or more of the hemipenis (which is barely bilobed), and have a strong tendency toward the posterior two teeth on the maxillary bone being distinctly enlarged compared to the teeth in front of them; the anterior maxillary teeth also may be distinctly enlarged. The species of *Lampropeltis* proper typically are diurnal (day-active) snakes that feed on mammals, while those of *Osceola* are nocturnal (night-active) snakes that feed mostly on lizards.

Gray-banded and Ruthven's Kingsnakes definitely belong to the *Osceola* group and could be considered either very primitive or very specialized milk snakes. Like the other species of the group (*L. triangulum, L. pyromelana, L. zonata*), *L. mexicana* and *L. ruthveni* are tricolored snakes bearing, at least in some specimens or populations, alternating black and pale (grayish to whitish) bands or saddles, the black bands having red centers (the black usually said to be split by red). In the Gray-band and Ruthven's the head is widened at the back and distinct from the neck, while in the other tricolors the head is relatively narrower and not distinct from the neck. (A word of caution is in order here, as some large *L. triangulum*, especially the tropical subspecies, have quite wide heads that to me, at least, appear quite distinct from the neck.) The real taxonomic characters of the species are largely to be found in scale counts, special attention being given to the rows of dorsal scales, number of ventral scales, and number of subcaudal scales. We'll go into these in more detail later, but basically *L. m. alterna* has 211 to 230 ventrals and 56 to 67 subcaudals; *L. m. mexicana* averages a few more rows, usually 23 to 27.

One odd feature of the Gray-band and Ruthven's is that at least in some (or most) specimens the belly is quite flabby to the touch. In typical kingsnakes, including milk snakes, the belly is heavily muscled and hard to the touch. Many Gray-banded Kingsnakes and Ruthven's Kingsnakes are soft-bellied and appear to lack the tight muscles

PHOTO: W. P. MARA.

Often considered one of the close relatives of the Gray-banded Kingsnake is the Sonoran Mountain Kingsnake, *L. pyromelana*, here represented by the popular Mexican subspecies *knoblochi*.

has 190 to 211 ventrals and 51 to 65 subcaudals; and *L. ruthveni* has 182 to 196 ventrals and 49 to 57 subcaudals. Notice that the number of scales increases from south to north. The two southern forms have 21 to 25 dorsal scale rows, while the northern *alterna* of the other species. Whether this is some type of illusion, a factor present only in captive-bred specimens, or represents a true difference in musculature of the abdomen is unknown to me. Recent studies of belly muscles in other snakes (the green snakes,

INTRODUCTION

Opheodrys, come to mind) have led to some interesting results. Similarly, a few differences in vertebral structure and skull proportions of these snakes have been noted in the literature. Because they were based on very small samples these differences may be meaningless, but certainly they deserve to be investigated further.

distinct species still is questioned by some biologists, including Dr. Kenneth Williams, who last revised the Milk Snake subspecies. However, as we will see later, *L. ruthveni* readily breeds with *L. mexicana* in captivity, and the two act like very close relatives.

For the last decade there has been a tendency to recognize two

PHOTO: K. H. SWITAK.

California Mountain Kingsnakes, *L. zonata* (here the subspecies *agalma*), are related to the Gray-bands but have pale (often red) snouts and many subtle pattern distinctions.

The first Gray-bands were described just over a century ago (1884) as a variety of the Milk Snake, *L. triangulum*, and until 1982 Ruthven's Kingsnake often was considered to be a synonym of one of the Mexican subspecies of Milk Snake, *L. t. arcifera*. There is little doubt that the three snakes are closely related, and the status of *L. ruthveni* as a

species in what is here called the Gray-banded Kingsnake, a northern Mexican and western Texas form with a higher ventral scale count and a tendency toward banding over the back (*L. alterna*) and a more southerly Mexican species with a lower ventral count and a tendency toward short blotches on the back (*L. mexicana*). However, there is

INTRODUCTION

PHOTOS: K. H. SWITAK.

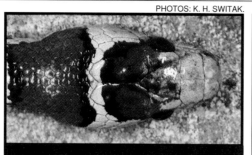

The whitish snout of the Sonoran Mountain Kingsnake (above, *L. pyromelana knoblochi*, below *L. p. pyromelana*) is distinctive. These milk snakes are very popular with breeders and are produced in fair numbers each year.

hard to pin down any characters that really would separate the two as full species. For our purposes in this book the Gray-banded Kingsnake is considered to have two subspecies, the northern true Gray-band, *L. mexicana alterna*, and the southern *L. m. mexicana*, often called the Mexican Kingsnake. As we will see, the other two forms often recognized by breeders, the Variable Kingsnake, *L. m. "thayeri,"* and the Durango Kingsnake, *L. m. "greeri,"* almost certainly represent intergrade populations between the two recognized subspecies. However, hobbyists maintain distinctive lineages or strains of snakes duplicating patterns similar to the type specimens of these synonyms and also obtain similar specimens as a result of various interbreeding experiments and selective breeding.

strong evidence that these two types of snakes interbreed where they have been found in contact in Mexico, and additionally it is

Gorgeous hybrids between albino Ruthven's Kingsnakes and the Blair's phase of the Gray-banded Kingsnake recently have been produced. Further breeding is expected to yield very evenly-patterned animals.

PHOTO: M. & J. WALLS, COURTESY D. RICHARD.

SOME HISTORY

Though the Gray-banded Kingsnake has been known for only a bit over a century, the species has suffered a more than normally complicated history. The following summary, though it may make a bit of tedious reading if you are not interested in taxonomy, attempts to follow the highlights of this history and present the major scientific references necessary to understand the present state of confusion in the species and also in Ruthven's Kingsnake.

1883: Our tangled history starts when Samuel Garman described two specimens of what he called *Ophibolus triangulus* var. *mexicanus* in his magnus opus on North American reptiles and amphibians (*Memoirs Museum of Comparative Zoology, Harvard*, 8(3): 66). (This volume sometimes is dated as 1884.) Unfortunately, Garman was somewhat out of his league when it came to a comprehensive coverage of the North American herps, and he only got as far as the snakes. Additionally, he seems to have had little imagination and was pretty much a taxonomic drudge, and few of his new names have survived to the present day. The two specimens were from near Ciudad San Luis Potosi, Mexico, in what today would be called the Saladan portion of the Chihuahuan Desert of north-central Mexico.

1893: In his volume on the reptiles and amphibians that appeared as part of the classic series *Biologia Centrali-Americana*, Albert C. Günther of the British Museum described a unique specimen of kingsnake under the name *Coronella leonis*. The single specimen had no other data than Nuevo Leon, Mexico, a large Mexican state. The specimen was well-illustrated and described, and because it was deposited in the British Museum it has been seen by several investigators since 1893, but it remains a truly odd specimen that never has been exactly duplicated. We'll discuss its characters in the chapter on the Mexican Kingsnake, but suffice it to say that it is a pale snake with small black-outlined red spots on the back, little or no pattern on the sides, and 197 ventral scales.

1897: Alfredo Dugès described a new genus and species of kingsnake, *Oreophis boulengeri*, from Guanajuato State, Mexico (*Proc. Zool. Soc. London*, 1897: 284-285). This genus was said to differ from *Ophibolus* (=*Lampropeltis*) in having the teeth at both front and back of the maxilla larger than those at the center, but later workers (Dunn, 1922, *Proc. Biol. Soc. Washington*, 35: 226) have recognized this to be an oddity of the type specimen, which obviously is a fairly typical specimen of *L. mexicana*, though with an unusually complicated black head pattern. The type has 185 ventral scales. The type specimen was redescribed and illustrated later by Hobart Smith.

1901: By the turn of the century, it was unexpected to find a new species of large snake in the United States, but Arthur Erwin Brown in that year described a new species, *Ophibolus alternus*, in the *Proc. Acad. Nat. Sci., Philadelphia*, 53(3): 612-613. The species had an unusual history, being represented only by a single specimen taken by a commercial collector (E. Meyenberg, who also collected the type of *Bogertophis subocularis*) and shipped alive to the Philadelphia Zoological Gardens (i.e., zoo). The specimen was said to come from the Davis Mountains of extreme western Texas, but for many years some workers felt that it was actually from Mexico and that such a distinctive species could not have been missed in well-collected Texas. The second specimen appears to have been collected in 1938 in the Chisos Mountains of Texas, and the third specimen was taken in Coahuila, Mexico, in 1939. By 1957 only seven specimen had been reported in

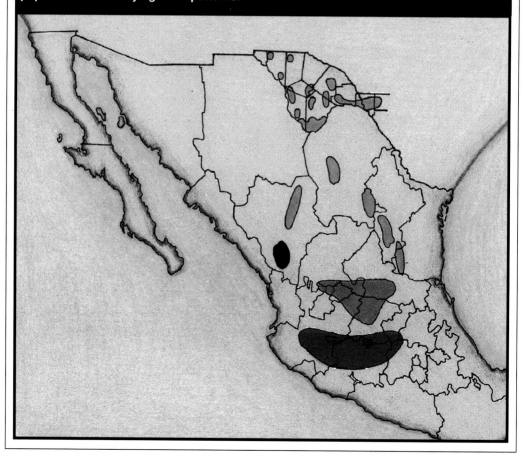

Distribution of the gray-banded kingsnakes in Mexico and western Texas. Green: *L. mexicana alterna*. Orange: *L. mexicana mexicana*. Red: *L. ruthveni*. The Greeri phase is indicated in black, while the Thayeri phase is indicated in blue; these are intergrade populations with varying color patterns.

SOME HISTORY

A typical Ruthven's Kingsnake, *L. ruthveni*. This species usually is seen as an albino today.

the scientific literature, but within the next decade collectors learned to hunt for it at night in appropriate habitats.

1920: Frank Blanchard, of the University of Michigan, described *Lampropeltis ruthveni* from a single specimen said to be from Patzcuaro, Michoacan, Mexico (*Occas. Papers Mus. Zool., Univ. Mich.*, No. 81: 8-10). Blanchard at this time was working hard on a revision of the kingsnakes and describing several new subspecies from the United States and Mexico. This particular specimen had been collected by the biological explorer Nelson in 1892, and some doubt has been cast on the exact collecting locality, though it probably was not far from Nelson's base camp at Patzcuaro. Blanchard described *ruthveni* as a full species but expressed doubts as to its relationships, a problem that has plagued later workers as well. Curiously, in the same paper he described *L. t. nelsoni*, one of the Milk Snakes with which *ruthveni* has been confused over the years.

1921: In this year Blanchard published his revision of the kingsnakes (*Bull. U. S. Natl. Mus.*, 114: 1-260) and presented full descriptions of each species and subspecies described to that time. He treated *L. mexicana, alterna,* and *ruthveni* as full species of uncertain relationships but probably related to the *triangulum* group, while *leonis* was postulated to be related to *L. calligaster*, a concept not accepted by any other workers. All the forms were illustrated from type specimens (except *leonis*). Blanchard's revision has today been partially supplanted by fuller reviews of each of the species of the genus, but it still is essential reading to understand the history of the group.

1924: Following from the stimulus of Blanchard's revision, Arthur Loveridge described *Lampropeltis thayeri* from a specimen from Miquihuana, Tamaulipas, Mexico (*Occas. Pap. Boston Soc. Nat. Hist.*, 5: 137-139). In many ways this specimen resembles the types of *L. mexicana*, but the red is applied in broad rings that extend to the tips of the ventral scales. Such ringed specimens are today selectively bred. In 1944, Hobart Smith (*Zool. Ser., Field Mus. Nat. Hist., Chicago*, 24: 135-152) described three other specimens that could be assigned to *L. thayeri*, though two of the three specimens (all juveniles) are clearly intermediate to *L. mexicana*; these specimens came from near Galeana, Nuevo Leon, Mexico.

1942: In an important paper in the *Proc. Rochester Acad. Sci.*, 8: 196-207, Hobart Smith discussed the relationships of the *Lampropeltis triangulum* group, including *L. mexicana* and *L. alterna*. He placed these two species plus *leonis* in a *mexicana* group, while the ringed *L. ruthveni* and *L. thayeri* were considered to be related to *L. pyromelana*. He also described the hemipenes of the various species and presented a key for the separation of Mexican species and subspecies of *Lampropeltis*, as well as redescribing and illustrating the type of *Oreophis boulengeri* (=*L. mexicana*).

1950: In *Copeia*, 1950(3): 215-217, Alvin Flury from the University of Texas described a new species of kingsnake, *L. blairi*, from a single male found dead on the road west of Dryden, Terrell Co., Texas, in 1948. Though considered obviously close to *L. alterna* and *L. mexicana*, it also was considered to be close to *L. doliata* (=*triangulum*). The only real difference was in the presence of 14 black-bordered red saddles on the body separated by 14 broad gray areas. In *L. alterna* narrow black rings alternate with black rings split by red. The second specimen of the new species was reported the following year (*Copeia*, 1951(4): 313) by Axtell, being a specimen from Val Verde Co., Texas. This specimen was illustrated in a nice black and white photo, and the note is accompanied by the first observations on a specimen of this group of snakes in captivity.

1957: The publication of Wright and Wright's *Handbook of Snakes of the United States and Canada* (Cornell Univ. Press) summarized what was known of the history and distribution of the rare *L. alterna* and *L. blairi* and presented black and white photos of each. Many of today's herpetologists spent hours poring over Wright and Wright as kids, noting especially the rare species like these.

1961: Robert G. Webb of the

SOME HISTORY

The holotype of *Lampropeltis greeri*, here considered the Greeri phase of *L. mexicana*.

University of Kansas stuck his neck out and described as a new species a juvenile male kingsnake caught in a mousetrap near Ciudad Durango, Durango, Mexico, in 1958 (*Copeia*, 1961(3): 326-333). Named *Lampropeltis greeri* in honor of his collecting partner (J. Keever Greer) when the snake was discovered, the oddly patterned specimen was illustrated and described in detail in an attempt to place it within the species related to *L. mexicana*. A key the six forms of the group was presented, all based on color patterns, and the holotype of *L. thayeri* was illustrated as well. *L. greeri* was considered to be very close to *L. alterna*, having narrow black cross-bands over the back, some split by red, but was said to differ by having a more complicated head pattern and all the bands complete across the back rather than some being broken into speckling. Today specimens from near Ciudad Durango are considered to represent an intergrade population between *L. alterna* and *L. mexicana*, one of the reasons the two forms are treated here as subspecies. A look at the illustration of the type specimen shows a head pattern much like

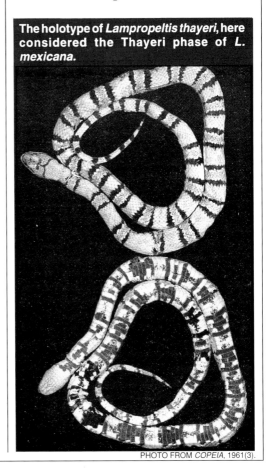

The holotype of *Lampropeltis thayeri*, here considered the Thayeri phase of *L. mexicana*.

that of *L. mexicana* and also some of the dorsal pattern is in the form of black-outlined dorsal blotches like that snake. There are 202 ventrals in the type specimen, well within the range for *L. mexicana* proper.

1962: If Webb stuck his neck out in 1961, Gehlbach and Baker chopped it off in 1962. In their paper in *Copeia*, 1962(2): 291-300, they reviewed basically all that was known of the snakes specimens than previous workers, and they were able to pinpoint several specimens or populations intermediate between the various forms. They presented a map of the distribution of the species and gave many notes on natural history, including feeding and activity period. *L. leonis* was considered to be a full species related to *mexicana* but was retained only because the pattern of the unique specimen was not

Until 1990 or so, Ruthven's Kingsnake was virtually unknown to scientists and hobbyists. Since then, brightly patterned albinos (here the phase with bright white and scarlet rings) have become popular though expensive pets.

related to *L. mexicana* and came to the conclusion that *alterna*, *blairi*, *greeri*, and *thayeri* were subspecies of *L. mexicana*. Their arguments were well-reasoned, they had seen may more duplicated by any known specimens of *mexicana*. In 1965, Gehlbach and McCoy (*Herpetologica*, 21(1): 35-38) described a few more specimens of the group and formally placed *greeri* in the synonymy of *L. m. mexicana* as being an intergrade population with *L. m. alterna*.

1967: In the *Catalogue of American Amphibians and Reptiles*, 35.1-35.2, Frederick Gehlbach reviewed the subspecies of *L. mexicana* and presented brief diagnoses and a detailed range map for the group. He recognized *L. m. mexicana*, *L. m. alterna*, *L. m. thayeri*, and *L. m. blairi* as valid subspecies but suggested that *blairi* and *alterna* could be color pattern polymorphs of a single subspecies and noted that *thayeri* was known from only four specimens, two of them intermediate to *L. m. mexicana*.

1970: Ernest C. Tanzer, in a paper in *Herpetologica*, 26(4): 419-428, reported that a wild-collected female of *L. m. alterna* from near Comstock, Val Verde Co., Texas, gave rise to a natural litter of six eggs, of which three hatched into *blairi* phases and two produced *alterna* phases (the other egg died). He considered this evidence that the two phases are just color pattern polymorphs, similar to the ringed and striped phases of *L. getula californiae* so familiar to hobbyists. The distribution of the two color phases is to some extent geographically separated (*alterna* west of the Marathon Basin, *blairi* east of the Basin), but there is considerable variation and many specimens intermediate between the two phases. Observations on the subspecies *L. m. alterna* in captivity are given. This appears to be the first record of the species reproducing in captivity, though not captive-bred. The first recorded captive-breds that I know of were reported by Murphy, Tryon, and Brecke in 1978 in *Herpetologica*, 34(1): 84-93, based on eggs laid in the Dallas and Fort Worth Zoos in various matings of the *alterna* and *blairi* phases. (A

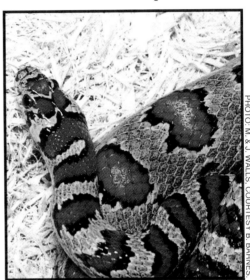

A typical adult Mexican Kingsnake (*L. m. mexicana*). This seems to be the form bred least commonly and the one in least demand, though it still is a beautiful snake by any standards.

paper by Wagner in a herp symposium in 1976 was only partially on Gray-bands.)

In 1979, Dennis J. Miller completed his Masters thesis, a study of the life history of the Gray-band in Texas. Published as *Chihuahuan Desert Research Inst. Contribution No. 87*, this paper examines in detail everything known about the Gray-band, with an analysis of variation in the snake and detailed range maps of basically all specimens known to that time. This paper remains the major review of *L. m. alterna*,

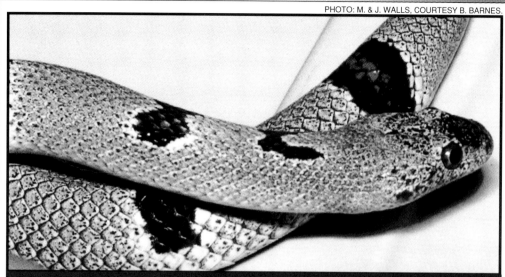

At one time the typical phase of the Gray-banded Kingsnake, *L. m. alterna*, was one of the most sought-after snakes in the U.S., but today it has been somewhat eclipsed by the more colorful Blair's phase. Dozens or perhaps hundreds still are taken from the wild each year in Texas and many more bred in captivity.

including its natural history, and emphasizes the variability of the snake in Texas.

1982: William R. Garstka, in a long paper published as *Breviora, Mus. Comp. Zool.*, No. 466: 1-35, reviewed the systematics of the *mexicana* species group and recognized three full species with no subspecies. He distinguished *alterna* as a full species based on small differences in eye color and vertebral shape from *mexicana* and removed *ruthveni* from the synonymy of *L. t. arcifera* where Williams had placed it in his 1978 review of the Milk Snake. Though the paper is very interesting, it generally is thought that Garstka viewed too few living specimens to back up his conclusions. He also fails to explain the numerous reports of intermediates between *alterna* and *mexicana*. Small differences in the spination of the hemipenes may be valid, but again they are based on examination of only a few specimens. Though his conclusions have been accepted by some workers, others are adamant that *alterna* and *mexicana* are intergrading subspecies, a view with which I am agreeing in this book. His comments on possible mimicry between members of this species and various small rattlesnakes with which they are found in nature are interesting and deserve more study.

As far as I know, there have been no other major papers on the *mexicana* group published since 1982, though the species have been treated often in the hobbyist literature after that date. Confusion still reigns and will continue to do so until modern studies based on cellular biochemistry and genetics have been conducted.

CAPTIVE CARE

Taking care of Gray-banded Kingsnakes and Ruthven's Kingsnakes is relatively simple. The animals are docile and undemanding, with only feeding being at all problematical. In fact, one of the reasons these snakes have become so popular is that it is possible to raise many specimens in a small area.

CAGES

Though adults of these two species commonly are 3 feet long, with a few giants reaching 4 feet and more, they do not need a large cage. Two types of setups commonly are used, a display terrarium and the shoe box method.

A good display terrarium is a 20-gallon all-glass aquarium or its equivalent (about 24 to 30 inches long and 12 inches wide) with a locking mesh lid. Because the snakes are nocturnal or at most crepuscular (active at dusk and dawn), it may be a good idea to cover three sides of the terrarium with construction paper to reduce the amount of light reaching the king. Regardless, it is best to keep the terrarium in a darkened area.

In the shoe box method, plastic boxes of various types and sizes are used as cages. These commonly are placed in racks that allow nine or a dozen or more boxes to be housed in just a few square feet of wall space. Shoe boxes are a great advantage if you are running (maintaining for breeding purposes) a dozen or more kingsnakes in a project. If the boxes fit tightly into the racks they do not need lids yet are easily pulled out to check the snake or perform routine maintenance. However, shoe boxes are lousy if you want to display a few specimens so your friends can look at your pets. Also, they always look cramped and the snakes seem unable to move freely, but this is not really the case. Remember that these

Gray-bands take little space and have very simple requirements. They often are kept in racks of plastic boxes to conserve space.

A simple setup for a Gray-band (here *L. m. mexicana*). These are among the easiest snakes to house.

kingsnakes spend the day in tight underground holes and crevices and don't move about much except at night when looking for food.

SUBSTRATE

Almost anything will work as bedding for a Gray-band or Ruthven's King as long as it does not hold much moisture. Clean newspaper is excellent and often is used in shoe boxes where it does not detract from the appearance. Many keepers use excelsior or even Spanish moss in shoe boxes because these may be cheap and easy to obtain in some areas. In the display terrarium a base of sand with some potting soil or peat moss added (no more than 50%) works well and looks fairly natural if many flat stones are layered over it.

HIDE BOXES

These are secretive snakes that usually hide all day (though they eventually learn to come out if the light levels are low), so they need lots of hiding places in order to feel secure. Commercial hide boxes are excellent, but be sure to buy one that is not too large—snakes like to have their backs against the wall to feel comfortable. Layers of rocks that form artificial crevices are fine; they should be anchored with silicone cement to each other and if necessary to the bottom of the terrarium.

OTHER DECORATIONS

Gray-bands and Ruthven's are not great climbers, but some do enjoy stretching out on a low branch above the bottom of the terrarium. Real plants will not do well in a kingsnake terrarium, but artificial plants look nice in the display terrarium and will not be noticed by the snakes.

HEATING AND LIGHT

Since these snakes do well at about 85°F and usually are comfortable at warm room temperatures, heating is not difficult. A heating tape or undertank heating pad placed at one end of the terrarium will provide all the heat needed in most climates. Shoe boxes generally are heated by a built-in heat tape toward the back of the

If you are looking for a substrate that is easy to work with, a terrarium liner will do nicely. Many pet shops now carry these in different sizes to fit many types of cages.

PHOTO COURTESY OF FOUR PAWS.

CAPTIVE CARE

rack. Remember that the heat source should be placed so the snake can take advantage of a thermal gradient, the cage ranging from warm at one end to relatively cool at the other. The desert and neighboring woodlands where these snakes are found are relatively cool at night when the snakes are most active.

instead sensing only changes in temperature over the seasons.

FEEDING

In nature, Gray-bands and probably Ruthven's feed mostly on lizards. They take many species of *Sceloporus*, swifts or spiny lizards, as well as other dry-land lizards in the 3- to 6-inch category. In

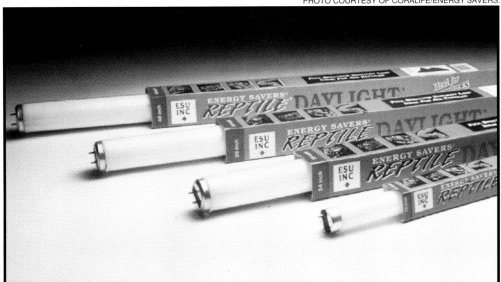

PHOTO COURTESY OF CORALIFE/ENERGY SAVERS.

A variety of light bulbs suitable for use in the terrarium can be found at your local pet shop.

Basking lights are not necessary, though sometimes specimens do become so adjusted to captivity that they will bask for an hour or two under a weak (25 to 50 watts) incandescent bulb. The full-spectrum fluorescent lights suggested for most snakes are not necessary for these kingsnakes and may make the snakes uncomfortable. Normal household light should be sufficient. There are strong indications that the snakes do not respond to photoperiod (length of daylight compared to night),

captivity they may take any lizard of appropriate size, including Mediterranean Geckos (*Hemidactylus turcicus*) and baby Green Anoles (*Anolis carolinensis*). If you are dealing with a wild-taken specimen, it may be necessary to have a small colony of food lizards or be able to purchase or collect specimens on a weekly basis. They also eat frogs, small mammals, and even reptile eggs in nature, but these cannot compare to their taste for lizards.

Because lizards are difficult to

Frozen and thawed mice will be taken by captive-bred Gray-banded and Ruthven's Kingsnakes. Today frozen mice are available at many pet shops.

supply on a regular basis, all Gray-bands and Ruthven's should be switched over to mice as soon as possible. Specimens over nine months old (about a third grown) should already be taking mice before you purchase them. With captive-bred adults this is no problem, but with many wild-caught specimens (still regularly available, perhaps unfortunately) it will be necessary to adapt the snakes to mice. The same also is true of young specimens, as we will discuss in the next chapter. Suffice it to say that starvation is the leading cause of death among baby Gray-banded Kingsnakes.

Two or three standard methods of adjusting to a mouse diet are used by keepers. A mouse (frozen and then thawed is best—live mice may cause damage) is rubbed with a dead lizard that is kept in the refrigerator for just such a use. Many keepers feel it is best to cut open the lizard and rub the mouse against the gut or to slit the skin and pass the mouse through the cut area. Lizard blood seems to be a strong stimulant to these kingsnakes.

Braining a small mouse (already humanely killed) also may work. The top of the mouse's head is cut open to expose the brains, which then are cut a bit. Still another method that works with small snakes is to feed them pieces of mouse tails.

Forced-feeding often is resorted to in these kingsnakes but is not a very good idea. It is most likely to work with young specimens, but these are even more delicate than adults. Adult Gray-bands are very docile and easily handled snakes, and it often is possible to

CAPTIVE CARE

simply insert a small mouse into the mouth of a non-feeding adult and have the snake swallow it.

Two fuzzies or small adult mice per week should be sufficient to maintain an adult snake's weight. More than that and the snake will become fat. Remember that these are not active snakes in captivity. Feed the animals in the dark or at least in subdued light that corresponds to their nocturnal hunting habits. Regurgitation of food a day or two after eating may indicate that the snake is too cool, the humidity is too high, there is too much light, or the snake has come down with a respiratory ailment and requires veterinary care.

Many breeders feel that a vitamin D3 supplement should be added to the diet about every second or third feeding. Reproduction rates may drop if there is no supplementation. A calcium supplement given to females of reproductive age may improve the texture of the egg shell, but the snakes should get sufficient calcium from their diet for normal functioning.

WATER

Though *mexicana* group kingsnakes are inhabitants of dry habitats, most really are not true desert animals. They must be given access to a bowl of clean water at all times, and they will

If this young albino Ruthven's Kingsnake takes the pinkie mouse, it probably will do well in captivity. Baby Ruthven's and Gray-bands are lizard-eaters in nature, a pattern that must be broken to make them successful terrarium animals.

PHOTO: I. FRANCAIS, COURTESY SHELTON.

drink often and well. Even hatchlings will drink within a day or two of leaving the egg. A water bowl must even be provided to adults in brumation.

AILMENTS

As usual with snake illnesses, there is not much that a hobbyist can do to either diagnose or treat internal ailments. All specimens should be wormed with an appropriate chemical (fenbendazole [Panacur] and Ivermectin often are recommended) under the supervision of a veterinarian who can adjust the dosage to the correct levels and also check the stool for parasites and their eggs. Wild-caught specimens may require a course of treatment that lasts several months, but captive-breds should be relatively free of parasites once they have switched from a lizard diet to mice.

Respiratory infections (harsh breathing, runny noses, frequent regurgitation) and mouth infections (cottony or cheesy material between teeth, swollen and distorted jaws) require immediate attention from your veterinarian.

Mites often become a problem with Gray-bands kept in display terraria with decorations that allow the pests to lay their eggs in protected areas. The best treatment, and the only one that can be recommended today, is sprays containing pyrethrum or synthetic pyrethrum derivatives. These plant-derived chemicals (and their laboratory-produced cognates) kill mites on contact without harming vertebrates. The snake is thoroughly wetted with the spray and then wiped dry. The terrarium must of course be

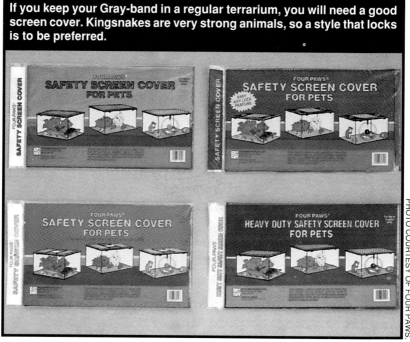

If you keep your Gray-band in a regular terrarium, you will need a good screen cover. Kingsnakes are very strong animals, so a style that locks is to be preferred.

CAPTIVE CARE

Like other kingsnakes, Gray-bands are constrictors, and they have stereotyped behavior that makes them go through the motions of constricting even dead mice before eating them. Photo: I. Francais, courtesy Shelton.

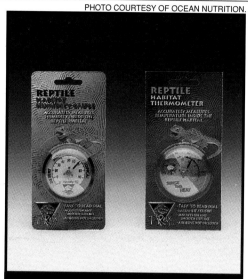

Every terrarium setup should have at least one thermometer, and it is a good idea to have a small hygrometer as well to measure the humidity.

ointment if necessary.

KEEPABILITY

Gray-banded Kingsnakes and Ruthven's Kingsnakes have many attractions, including their extreme docility. Even specimens encountered in the wild seldom struggle when picked up and typically do not musk or bite. Captive-bred specimens are gentle, non-aggressive snakes that can be kept in pairs or small groups without friction (except during mating season). As with all other snakes, however, it is best not to keep specimens of different sizes together, especially hatchlings and adults. Once a young specimen has begun to feed on mice as the regular diet, it may live for 15 years or more. Though the cost of purchasing a specimen may be high, they are well worth the money, especially if you can get a specimen that is a proven mouse-eater.

Well-fed specimens may be able to breed when about 24 months old, and today it is quite possible to actually recoup the money you paid for a small group of young Gray-bands or Ruthven's after one or two successful breedings. However, these snakes are being bred in increasingly larger numbers each year by more and more people, so it is likely that prices will drop significantly in the near future. These are beautiful snakes, and you are much better buying them as pets rather than investments, but in the next chapter we'll discuss the basics of breeding anyway because that is the goal of most keepers.

completely cleaned, all decorations sanitized, and the substrate changed before the treated snake is returned to the terrarium. Very small snakes should be treated with great care because there have been a few reports of deaths among hatchlings treated with pyrethrum sprays.

Ticks are likely to be found only on wild-caught specimens. They are small dark parasites usually found under the edges of the scales and often near the end of the tail. Though not a great problem in these kingsnakes, they should be removed by dabbing with alcohol and then pulling them out with fine forceps by using a slight twisting motion to help loosen the mouthparts. Keep an eye out for infection at the sites of tick removals and treat it with a commercial iodine

BREEDING

Though once difficult snakes to breed, Gray-bands and to some extent Ruthven's have become relatively simple to breed through several generations. Today the magazines are filled with ads for breeders of these species, including locality-specific lineages of Gray-bands and albino Ruthven's Kingsnakes. Though they still cannot be considered beginner's snakes because of complications involved in getting hatchlings to feed, they have become rather common among more advanced snake keepers and prices have dropped to some extent. Though yearling and adult specimens are much more expensive than hatchlings that have not yet fed, they probably are a better investment because they have a better longevity record.

SEXES

Though some breeders feel that they can sex Gray-bands and Ruthven's by probing, many others agree that this standard method of sexing is unreliable in these kingsnakes. The hemipenes of *mexicana* kingsnakes are rather short, extending for only some seven to ten subcaudal scales beyond the cloaca.

Males when young can be sexed with some accuracy by carefully forcing them to erect a hemipene. The snake is held firmly coiled around one hand, while the thumb and first finger of the other hand are used to carefully roll

Gray-banded Kingsnakes are difficult to sex by appearance. Many keepers prefer to "pop" the hemipenes of young specimens (before the muscles become stronger) and keep records of sexes this way. Popping hemipenes can be dangerous to the snake, so it should be done under supervision.

along one side of the tail, moving from the tip toward the cloaca. This motion often is sufficient to force the muscles that hold the hemipene inverted in its pouch to erect the hemipene at least partially from the cloaca. There are two main drawbacks to this technique. First, if too much pressure is exerted, it is possible to damage the hemipene, its muscles, and/or the sheath, leading to future mating difficulties. Second, older males

Probing does not always provide reliable sex determination in these kingsnakes. Many males probe to a depth of only seven subcaudals, while females may have anal pouches that are four or five subcaudals deep, leading to a strong chance of error.

have stronger muscles and can resist delicate pressure without erecting a hemipene; if the pressure is increased, the chances of damage also are increased. The erection of a hemipene is positive proof that the specimen is a male, but if no hemipene is erected it is not proof that the specimen is a female—it might be a strong male.

Males often have wider, more muscular heads than females, or at least so some breeders believe. Adult males often are smaller and more slender than adult females, but there is so much variation in specimens that this is a very hard factor to judge. Males are more combative than females, especially if two males are placed in the same cage at the beginning of the breeding season.

BRUMATION

Successful breeding of *mexicana* group kingsnakes seems to require a cooling period of about three months at about 55°F. Experiments by several breeders have shown that altering the day length (photoperiod) has little or no effect on inducing breeding, so fortunately it is not necessary to track day lengths for these species.

Make sure your breeding specimens are fully adult and capable of breeding. Minimal age of breeding seems to be about 18 months, but the best chance of getting fertile eggs is by mating

BREEDING

specimens at least 24 to 30 months old. Mating success seems to increase if several specimens, including perhaps two males and two to four females, are kept together as a mating group during brumation and at the very beginning of the mating season next spring. Sex the animals as best you can and go with numbers to increase your odds of success. The animals should have been fed well the previous summer and autumn, be healthy and without worms, and not have external parasites.

In mid-November begin to reduce the feeding interval and amounts. The snakes must have empty guts when they go into the cooling period. Also gradually drop the temperature of the cage, perhaps lowering it by about six degrees each week. By the middle of December the cage temperature should be 55°F, the guts of the snakes should be empty, and the snakes should be inactive. Brumating (partial hibernation) kingsnakes can tolerate temperatures as low as 50°F without harm and become active when the temperature rises to 60°F. Keep the cage dry, with all the usual hiding places, but be sure a water bowl is available for the short periods when the snakes might become active.

Hold the snakes in brumation from about mid-December through mid-March, at which time the temperature should be

Products of the breeder's art: a brilliant Blair's phase Gray-band and an albino Ruthven's. The Ruthven's has a greatly reduced red pattern that some keepers (but certainly not all) would consider a major fault, but the *"blairi"* is well-marked and has a dark head preferred by some keepers.

PHOTO: M. & J. WALLS, COURTESY B. BARNES.

gradually raised to about 70°F. At this temperature the snakes become almost fully active and will begin to accept mice as usual. Continue slowly raising the temperature to the normal 80 to 85°F by the beginning of April.

The snakes should be separated as they come out of brumation, but by the time they are feeding and the temperature is back to normal they should be ready to breed. You might want to put two males together at first to initiate combat before mating or just to check the sexing. As a general rule, females are put into the male's cage for mating, but these snakes really are not too choosy about such details.

MATING

When a willing female is placed in the cage with an adult male, the male usually shows interest almost immediately. However, since mating often occurs at night, the pair should be allowed to remain together for several weeks if necessary if mating is not seen sooner. Putting a second male in the cage with the pair usually leads to combat between the males. This consists of the males moving alongside each other and forming arches with their bodies as well as attempting to entwine each other. The purpose of the battle, which usually is non-violent, is for one male to turn the other over so its back is against the ground.

A breeding pair of Mexican Kingsnakes, *L. m. mexicana*. The large difference in color patterns does not interfere with mating ability.

BREEDING

A breeding pair of Thayeri phase Gray-banded Kingsnakes. Notice the way the tails are coiled about each other and the position of the male's head behind the female's.

Combat may last just a few minutes or as much as five to six hours if one male is not removed sooner. Many breeders think that male-male combat promotes breeding by the stronger male, but it also stresses both males.

If a second male is present while a pair is trying to mate, he may interrupt breeding behavior of the pair, though such interruptions usually are of short duration and probably only increase the speed with which the pair mate.

A male moves alongside a female, constantly flicking his tongue and trying to get his body parallel to hers. He moves along her side and back, using his body to force her to stop. If the female is willing to mate, she signals this by raising her tail a bit. The male moves his tail parallel to hers but does not wrap around her tail. He inserts a hemipenis from alongside the cloaca rather than under it. Copulation may last only a few minutes or be somewhat more prolonged. After mating the snakes usually slowly separate and go their own ways. Disengagement may produce a bit of bloody fluid from both the female's cloaca and the male's hemipenis, but it is nothing to worry about.

EGG-LAYING TO HATCHING

Mating typically occurs between April and June, and the eggs are laid about a month later. Females often prefer to lay in a dish of moist sphagnum. The clutches range from 5 to 12 eggs, seldom more. Typical eggs are about 1.25 to 1.5 inches long, oval, and white with a fairly thick shell.

If incubated in moist vermiculite (one part water to two parts vermiculite) at about 82°F, incubation time varies between 50 and 70 days, sometimes a bit longer. The babies cut through the shell but stay inside for one to two days while absorbing the yolk and learning to breathe. Young are about 8 to 10 inches long on emergence from the egg. In a week to ten days they have their first shed and are ready to take their first meal. Provide them with a shallow dish of water from the beginning.

FEEDING YOUNG

At this point the major problem with *mexicana* group kingsnakes comes to the fore. In nature

A deformed egg or slug produced by *L. m. alterna*. Such infertile eggs are not uncommon in most clutches produced in captivity.

PHOTO: M. & J. WALLS, COURTESY B. BARNES.

BREEDING 31

Just before copulation begins, the female kingsnake lifts her tail to expose the cloaca, the final sign for the male that she is ready to breed. Photo of Thayeri phase snakes.

PHOTO: I. FRANCAIS, COURTESY SHELTON.

hatchlings feed on small lizards, probably mostly *Sceloporus* (swifts) and *Cnemidophorus* (whiptails) taken at night while they are sleeping. So far, captive-breds attempt to maintain their natural feeding habits, and they want lizards. Small geckos may be taken, but you will need lizards of some type at first.

However, no one wants to continue to feed lizards to kingsnakes, and everyone attempts to convert hatchlings to pinkie mice as soon as possible. This creates the second problem, for as many as half of all hatchlings will not feed on mice and die. Additionally, many hatchlings will not feed on available lizards in captivity either. To successfully mature, a young *mexicana* group kingsnake must be feeding on mice by the time it is six months old.

The usual ways to get a king to eat pinkie mice include scenting the mouse by rubbing it with a lizard, slitting open the skull so the brain is exposed, and feeding bits of mouse tail. If you are lucky, however, placing a live pinkie in a darkened container with the snake and then putting the container back into the snake's cage overnight may work.

In extreme situations, forced-feeding may be necessary, but the small size and rather delicate nature of the young make this difficult. Attempts to force open the mouth and stuff pieces of pinkie (such as a hind leg) often result in regurgitation. Dicing a pinkie in a small amount of water and feeding it to the snake through a plastic tube attached to a syringe may work better but also can kill the snake if done improperly. A pinkie pump, which basically is a syringe with blades on the plunger that mince a pinkie within the tube so it can be forced through a plastic tube into the snake's stomach, is an expensive piece of equipment that is used in desperation. All methods of forced-feeding are dangerous and must be learned by seeing an experienced keeper do them first. Young Gray-bands and Ruthven's are too expensive to experiment on, and you wouldn't want to accidentally kill the results of six months' work (from brumation through hatching).

The high losses among babies perhaps are the major reason these kingsnakes have not yet replaced the Corn Snake as the most common colubrid snake in captivity. As more and more clutches are hatched each year, however, it is likely that selection will weed out whatever behavioral genes restrict young Gray-bands to a lizard diet, and mouse-eaters should become more common and easier to convert from lizard diets. When this happens the Gray-band will become even more common and popular in the terrarium hobby.

Kingsnake eggs are broadly oval and usually creamy white within a few hours of laying, but they may darken considerably during development and develop brown spots. Most breeders use moistened vermiculite as an incubator substrate.

BREEDING

PHOTO: I. FRANCAIS, COURTESY SHELTON.

RUTHVEN'S KINGSNAKES

HISTORY

How do you determine the relationships of a snake that is very rare in nature (or at least very difficult to collect) and closely resembles in pattern and structure several other known species? Well, usually you make a guess. When Frank Blanchard described *Lampropeltis ruthveni* (named after Alexander Ruthven, herpetologist and academician at the University of Michigan) in 1920, he had a single specimen that was collected in 1892 by E. W. Nelson, noted biologist and explorer in Mexico. Though the type locality was given as Patzcuaro, Michoacan, Mexico, that may represent the base camp of the collector rather than the actual locality; supposedly the type specimen now bears a locality label reading "Potrenaro." Based on his experience with other central Mexican kingsnakes, mostly subspecies of *L. triangulum*, Blanchard decided that the unique specimen represented a full species of uncertain relationships. He found it easy to distinguish from other red-banded kings of the area because it had: 1) a solid black

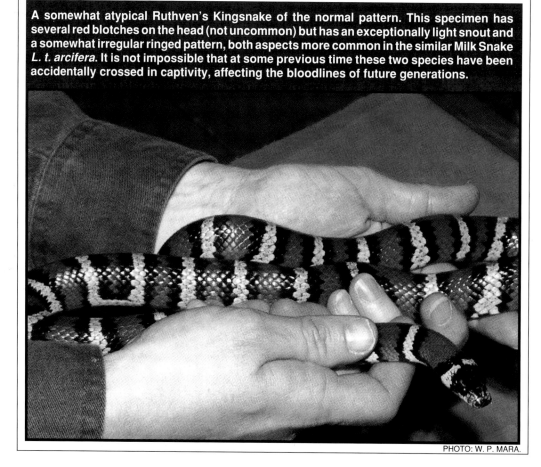

A somewhat atypical Ruthven's Kingsnake of the normal pattern. This specimen has several red blotches on the head (not uncommon) but has an exceptionally light snout and a somewhat irregular ringed pattern, both aspects more common in the similar Milk Snake *L. t. arcifera*. It is not impossible that at some previous time these two species have been accidentally crossed in captivity, affecting the bloodlines of future generations.

PHOTO: W. P. MARA.

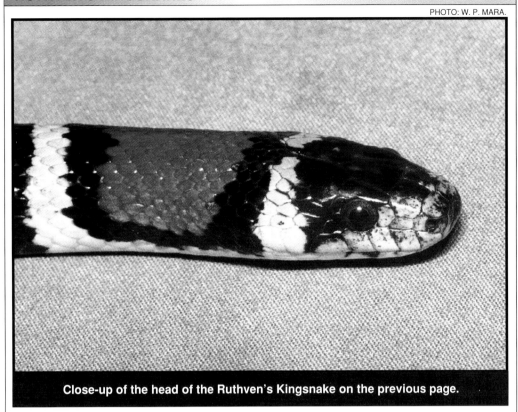
Close-up of the head of the Ruthven's Kingsnake on the previous page.

head (no white on snout); 2) no black tips to the scales in the red bands; 3) no tendency for the black rings to put out extensions through the red toward the next black ring. In other words, this was a very simply and cleanly ringed kingsnake with the standard white-black-red-black-white pattern. Additionally, the ventral scale count was low, only 189 plates, and the total number of white rings on the body (24) plus tail (6) equaled only 30.

By 1940 the species was known from several localities in central Michoacan, but its relationships with the *mexicana* group were still not suspected. In fact, it and the nearly ringed *thayeri* from eastern Mexico were thought to be relatives of *L. pyromelana* and *L. zonata*, which have somewhat similar patterns. In 1978, Ken Williams thought that *ruthveni* might represent a color phase of *L. triangulum arcifera* from the same region of Mexico.

In 1982, Garstka made the jump in logic that put *L. ruthveni* into the *mexicana* group. He worked largely on the basis that at some localities *ruthveni* can be collected along with *arcifera* or another subspecies of *triangulum*. Since there are differences in structure (scale counts) between the sympatric *ruthveni* and *triangulum* forms, they cannot represent color phases or polymorphisms as appears to be the case with *alterna*—*blairi* in Texas. Similarities to the *thayeri* phase of *L. mexicana mexicana* led to the deduction that *ruthveni* represents a distinct species related to *mexicana*. Hobbyists now have shown that *ruthveni* readily mates with *alterna* in captivity and produces normal offspring that generally resemble *alterna* in color pattern; I am not aware of fertile hybrids between *ruthveni* and any *triangulum* subspecies, but I doubt that they are impossible.

As more specimens of *L. ruthveni* became known to herpetologists, it was only a matter of time until hobbyists got a few specimens and started breeding them. It would seem that most or all the original hobby stock of the species came from the vicinity of La Piedad, Querétaro, Mexico, where the species is (or was) rather common and easy to collect. This seems to be the reason that the species often is carried on lists as the Queretaro Kingsnake. Since the range of the species is quite a bit broader, this name seems inappropriate to me and I prefer to stick with the more familiar Ruthven's Kingsnake.

DESCRIPTION

Typical Ruthven's are rather slender kingsnakes that reach a length of 30 to 32 inches, occasionally a bit longer. (There is a record of an old captive male over 40 inches long.) The head is not especially expanded in the temporal area but is moderately distinct from the neck. The hemipenis has claw-shaped

An old male Ruthven's Kingsnake with the pale rings toned with dark and a fairly obvious pale snout.

spines about 0.6 mm long. The tail is relatively long, about 15% of the total length. Scale counts of importance include: ventrals 182-196; subcaudals 49-57; dorsal scale rows 21-25; supralabials 6-8; infralabials 8-9. The color pattern is very simple, consisting of about 24 white rings about 2 scales wide on the body. These separate triads (black-red-black) consisting of black rings about 2 scales wide bounding a red ring typically 3 scales wide. The black rings are supposed to be narrowly edged with pale greenish in wild-caught specimens, but if this color is present in captive-breds I certainly cannot distinguish it. The rings extend around the body but tend to be irregular on the belly (which may appear mottled with red, black, and white), and there often are tinges of tan on the lower sides in the white rings.

The entire top of the head is black, occasionally with a few passes through the white ring, looking much like one of the many head patterns found in *L. m. mexicana*. The red color is bright and clean, without black tips or any extensions of the black rings into them. Typically all the rings have very sharp edges, not the jagged, irregular margins often seen in various *L. triangulum* subspecies The eye generally is brownish in color.

Though, as mentioned, the rings generally are sharp and uniform, in many specimens there are abnormalities of the rings. The

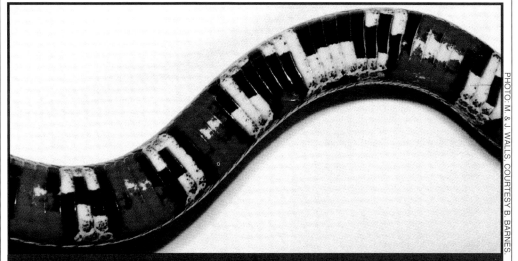

The belly pattern of the male Ruthven's Kingsnake shown on the previous page. Belly patterns in this kingsnake and the Gray-banded are extremely variable.

small red blotches, and the sides of the head also are black except for small white specks on the lip scales. The black ends cleanly at the back of the head, where it is followed by a narrow white ring and then the first black ring. Occasionally (at least in captive-bred specimens) the black of the head is connected to the first black ring by a median spur that two most common that I have seen are two rings connected across the back by expanded red, eliminating the two black and one white rings across the top, and closing the ring at the bottom by connecting of the black rings below a red ring to produce a long, wide blotch. This latter variation produces a pattern that closely resembles the pattern of

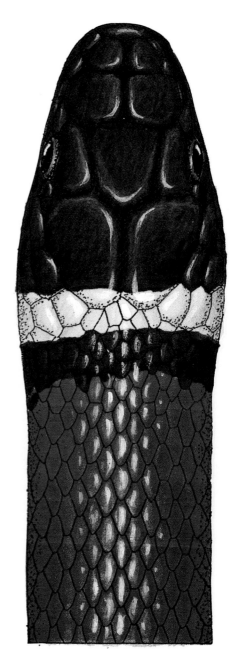

Comparison of the head patterns of *L. ruthveni* (left) and *L. triangulum arcifera*. Art: J. R. Quinn.

RUTHVEN'S KINGSNAKES

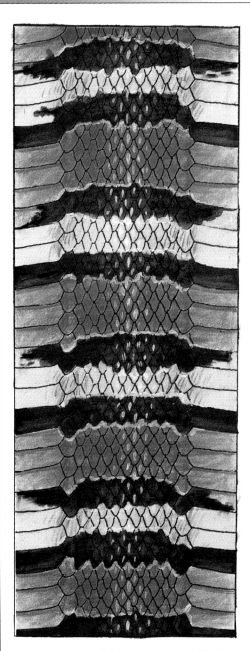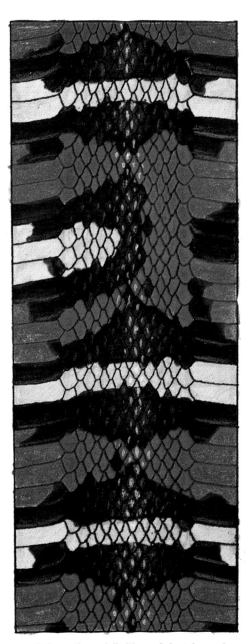

Comparison of fairly average midbody patterns of *L. ruthveni* (left) and *L. triangulum arcifera*. The extension of the black rings toward each other along the midback is common in the Milk Snake but rare in Ruthven's. Colors and details of pattern in both forms are broadly variable. Art: J. R. Quinn.

some *L. m. mexicana* and especially some *thayeri* phase specimens.

ALBINOS

The majority of specimens of Ruthven's Kingsnake sold today probably are albinos or specimens heterozygous for albinism. So far this is the only *mexicana* group king that is commonly available as an albino. The mutation is a simple genetic recessive, thus when two albinos (yy) are bred all their offspring will be albinos. If an albino (yy) is bred to a normal adult (YY), all the offspring will be heterozygous for albinism (Yy). These hetero specimens are indistinguishable externally from normal (YY) specimens, but if bred together (Yy X Yy), they produce one normal (YY) to two heteros (Yy) and one albino (yy). Heteros sell for about 20% the price of albinos, so they are a great and rather cheap way to introduce albinism into a line of snakes.

There seem to be several different shades or lines of albinos in the hobby, some breeders producing one type, others a different one. Whether these represent different mutations (unlikely) or specimens selected for their coloration is uncertain. Specimens are valued by the evenness of their pale coloration and brightness of the red. In one type of albino the formerly black rings are distinctly yellow and

Belly pattern of the new albino Ruthven's by Blair's phase Gray-banded Kingsnakes. This hybrid combination has produced great regularity of pattern in the offspring, both albino and heterozygous, resulting in very evenly marked snakes favored by enthusiasts.

PHOTO: M. & J. WALLS, COURTESY D. RICHARD.

An albino Ruthven's Kingsnake with scarlet and bright nacreous white rings. In some selected lines of this albino the formerly black rings are white, not pink. An unmarked head is preferred by most keepers of these snakes.

bound a bright white ring; the red of this type may be distinctly orangish. In another basic type the formerly black rings are white and hard to distinguish from the white ring they bound; the red ring of this type may be bright scarlet (coral) red. Collectors tend to prefer the second type to the first type. In all the albinos the head is white to yellowish with or without small red blotches and specks. Specimens with clean heads are valued more than specked heads. Obviously there can be many combinations of color shade in these albinos, but all are expensive.

Like normals, albinos are gentle snakes and can be kept like ordinary specimens. Both albinos and normals will mate with *alterna* and produce viable offspring, though obviously it takes much work over several generations to introduce the albino gene into an *alterna* or *blairi* phase snake. Recently just such albino hybrids have reached the market in limited numbers. Albino Ruthven's by *blairi* phase specimens are strikingly beautiful snakes. The heteros (heterozygous for albinism) have very cleanly edged bands and may be studies in symmetry.

REPRODUCTION

Ruthven's Kingsnake can be kept much like other *mexicana* types and also breeds like them. You might want to keep them a bit warmer than most Gray-bands and give them a weak basking light. Healthy adults must be brumated for about three months at 55°F, bringing them out to normal temperatures and feeding habits between March and April. Mating is as for Gray-bands, with the eggs being laid about a month after mating, usually in late May or early June. Clutches run about five to ten eggs, with about 75% hatching. As is to be expected, older (30 months plus), larger females lay more eggs and have a somewhat higher success rate than females that have barely turned two years of age. A second clutch of fewer eggs may be laid in July or August (double-clutching). Expect a good female to produce about eight to perhaps ten young per year.

Like Gray-bands, this species may show a decreasing fertility after two or three years of successful breeding. The reason for this is unknown, but remember that in nature these snakes feed on lizards and we have switched them to an unnatural mouse diet. There likely are differences in vitamins and other nutritional factors in a mouse versus lizard diet.

The incubation period for Ruthven's is short, about 50 to 70 days, often closer to the 50 than the 70. Incubation temperatures may range between 82°F and 85°F. The young are just as hard to get eating as are Gray-bands,

A different line of albino Ruthven's Kingsnake. Here the white and formerly black rings are difficult to distinguish, though in this case the entire pale pattern has a faint yellowish tinge. The red bands are somewhat uneven in both width and outline, a fault among some keepers.

PHOTO: M. & J. WALLS, COURTESY B. BARNES.

RUTHVEN'S KINGSNAKES

Today more albino Ruthven's Kingsnakes, and specimens heterozygous for albino, are sold than are normal specimens. Perhaps this is partially because normal Ruthven's are difficult to distinguish from Milk Snakes.

and they have even smaller heads and may be more delicate. Some keepers, however, have good luck feeding pinkies right from the beginning if each baby is kept in a separate small cage that is dark and well-ventilated and has a tight hiding spot. If possible, try not to start by feeding lizards and go directly to mice even if you have to risk forced-feeding.

DISTRIBUTION

Presently *L. ruthveni* is known from a broad arc through Jalisco, Michoacan, and Querétaro states in western central Mexico northwest of Mexico City. There it is found in dry, rocky woodlands at relatively high altitudes. This range is just to the south of the range of *L. mexicana mexicana*. So far no intergrades between *L. ruthveni* and *L. mexicana* have been reported, but they might be expected to occur.

THE FUTURE

Albinos are among the most striking snakes available in the hobby today, and they have perhaps unfortunately overshadowed the normal but brightly colored snakes. However, many hobbyists have trouble telling normals from Milk Snakes and question high prices for what seems to be just another tricolored king. Prices are dropping on both normals and albinos, but the usual problem of difficult to feed young with high mortality rates is a serious disadvantage. Adults are gorgeous, docile, easy to handle snakes that are a pleasure to own.

MEXICAN KINGSNAKES

As a general rule, I personally do not like to assign common names to subspecies. I feel that common names let many hobbyists equate species and subspecies because both have similarly formed names, thus defeating the various rules that let scientists always distinguish between the two groups. In some cases, however, tradition and market values force an exception. In this book the species *Lampropeltis mexicana* is being called the Gray-banded Kingsnake because that common name has a type of priority and has been widely used in the technical literature. This species is considered to have just two recognizable subspecies (in addition to two or perhaps three color phases) that are here called: *L. mexicana mexicana*, the Mexican Kingsnake, and *L. mexicana alterna*, the true Gray-banded Kingsnake. This may be confusing, and I am tempted to supply new common names for everything, but certainly hobbyists would object. Thus we are stuck with these common names. Some literature, by the way, refers to *L. mexicana mexicana* as the San Luis Potosi Kingsnake in reference to the type locality.

HISTORY

The history of this form, the first-described or nominate subspecies, is fairly short. It was described by Samuel Garman in 1883 from two specimens in the Museum of Comparative Zoology that had been collected near San Luis Potosi, in central Mexico. When Blanchard revised the kingsnakes in 1921, he knew only the types, which he redescribed in some detail; he also presented an excellent scalation drawing showing the midbody pattern. Twenty years later Hobart Smith redescribed and illustrated the type specimen of *Oreophis boulengeri* Dugès, 1897, confirmed that it was indeed the same as Garman's *mexicana*, and discussed a series of the form (then considered a full species without subspecies) from Alvarez,

Close-up of the head of the Mexican Kingsnake, *L. mexicana mexicana*. The blotch on the nape is set well forward and contacts or nearly contacts the usually paired spots on the back of the head.

PHOTO: M. & J. WALLS, COURTESY B. BARNES.

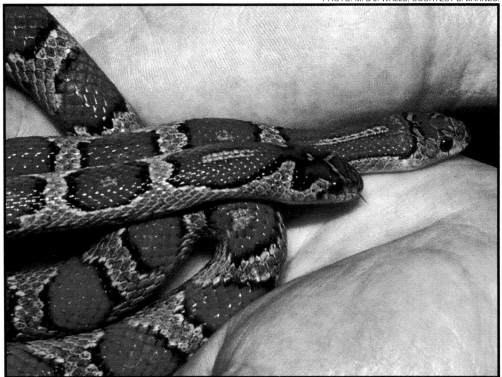

Two well-marked and fairly typical Mexican Kingsnakes. Notice both the head pattern and the squareness of the red saddles, which usually have the black edges reduced at the sides.

San Luis Potosi. A few years later he and Edward Taylor recorded the species from central Guanajuato, Mexico.

In the 1960's knowledge of this form grew rapidly as collecting crews from various American universities collected heavily in Mexico. In 1961, Webb described *L. greeri* from Durango in western Mexico, a form combined the next year with *L. mexicana* by Gehlbach and Baker. They also reduced all the described forms of the *mexicana* group then recognized as subspecies or synonyms of *L. mexicana*, including *greeri*, *thayeri*, *alterna*, and *blairi*. Gehlbach and McCoy in 1965 expanded on these relationships and considered *greeri* to represent an intergrade between *L. m. mexicana* and *L. m. alterna*, and thus a synonym.

This nomenclature was in use until 1982, when Garstka removed *alterna* (and its color form *blairi*) from *L. mexicana* and made it a full species again. As we have seen, this relationship generally is not accepted by herpetologists. He also treated *greeri* and *thayeri* as simple synonyms of *mexicana*, along with the mystery snake, *leonis*.

In this chapter I will try to describe the major color varieties or phases of *L. mexicana mexicana*, including the form still called *thayeri* by hobbyists. It must be understood that in nature these snakes seldom are collected and the specimens in captivity derive from just a handful of original specimens. Thus the captive-bred Mexican Kingsnakes today represent forms that may not accurately reflect the subspecies as it exists in nature. The various color phases have been interbred among themselves and *alterna*, and possibly even with *triangulum* subspecies, and their gene pool is far from pure. When selection for bright colors and adaptability to captive conditions is added to this melange, it must be considered that all "*mexicana*" and possibly many *alterna* in captivity really are, technically, mongrels.

DESCRIPTION

Typical Mexican Kingsnakes from central Mexico are slender kingsnakes with distinctly widened heads and an obvious neck. The tail is relatively long, about 15% of the total length, and often is demarked by an expanded red blotch above its base that extends well onto the cloacal area. Adults are about 24 to 32 inches long, with a few odd specimens reaching or exceeding 36 inches. The hemipenis has short (0.4 mm long), nearly straight spines. Important scale counts include: ventrals 190-200; subcaudals 51-65; dorsal scale rows 21-25; supralabials 7-8; infralabials 8-11. In this typical form the color pattern is simple and relatively constant. Against a whitish gray background there are 30 to 47 (typically about 35) dorsal blotches. These are bright red to dull reddish brown, about 3 to 5

Diagrammatic representations of the head and midbody patterns of typical *L. mexicana mexicana* (to the right) and the Greeri phase (to the left) representing variable intergrades of *L. m. mexicana* with *L. m. alterna* in western Mexico.

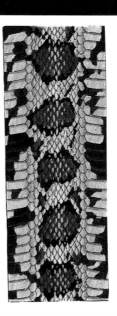

ART: J. R. QUINN.

MEXICAN KINGSNAKES 47

The illustration of the holotype of *Coronella leonis* (now *Lampropeltis mexicana*) published by Guenther in 1893. In many respects this appears to be an aberrant individual with characters of both *alterna* and typical *mexicana*, which might be expected from its collection locality in Nuevo Leon, Mexico, more or less between the ranges of the two subspecies.

scales wide at their widest, and are surrounded by black lines about 1 to 2 scales wide. The whitish areas between the blotches are narrower than the blotches, often only 2 or 3 scales wide. In most specimens the blotches extend downward on each side to several scale rows above the ventral scales (commonly 2 to 5 rows) and the black outlining becomes weaker near the ventral edges. There often are small black or black and red blotches low on the side between the major blotches.

The head pattern is complicated, usually consisting of a large black-outlined blotch on the nape of the neck that is connected by a line or spur to pairs of black or black and red blotches or black lines on top of the head. Often the nuchal blotch has a pale center. Commonly there is a broad black band back from the eye toward the angle of the jaws. The red color of the body blotches is clean, without black tips on the included scales. The belly varies from nearly white with only traces of small black spots to heavily mottled or partially banded with red, white, and black. In most specimens the black outlining of the red dorsal blotches is itself outlined with a narrow whitish to pale greenish line that contrasts with the grayish white background color. The blotches may be very irregular but always are numerous compared to the other subspecies and color phases. The eyes are brownish.

This typical form is found from southern Nuevo Leon through San Luis Potosi and Guanajuato, then north through Zacatecas into southern Durango, where it intergrades with *L. m. alterna*. It intergrades or at least varies into the *thayeri* form and possibly *blairi* in southern Nuevo Leon and possibly adjacent Tamaulipas.

THE *THAYERI* COLOR PHASE

The ringed *mexicana* color phase known as *thayeri* was described by Loveridge in 1924 from southern Tamaulipas, Mexico. About 20 years later three other specimens were described from near Galeana, Nuevo Leon, Mexico, but two of the three appear to be intergrades with typical *mexicana* or perhaps even with *alterna* (*blairi*). As originally described, *thayeri* is a ringed Mexican Kingsnake. The red

PHOTOS: M. & J. WALLS, COURTESY B. BARNES (TOP) AND R. DEUEL (BOTTOM).

Contrast the typical specimen of *L. mexicana mexicana* shown above with the rather brightly marked young specimen of the Greeri phase or Durango Mountain Kingsnake shown below. Durango Kings have been selectively bred in captivity and may today represent a totally unpredictable mixture of genes.

dorsal blotches extend at least to the first scale row low on the sides and often extend completely around the body. They are outlined by black as in typical *mexicana*, but the black may not extend all the way to the ventral scales along with the red. The belly usually is red and white with scattered black blotches that may be numerous or few. The nuchal blotch is red, usually lacks a pale center, and is quite long. It connects to the red head by a

MEXICAN KINGSNAKES

Two extremes of the Durango Mountain Kingsnake, an intergrade between *L. m. mexicana* and *L. m. alterna*. Probably no two specimens of this snake are the same, and some dealers use the name *greeri* for any oddball pattern. The small snake in the bottom photo bears a strong resemblance to the type of *leonis*.

Close-up of the head of a blackish kingsnake sold as a melanistic Thayeri phase *L. m. mexicana*. You will have to trust your dealer for identification of such blackish snakes because they lack patterns and thus cannot be identified to subspecies or phase. Melanistic Thayeri phase snakes are bred and sold in fair numbers.

In the Thayeri phase of *L. mexicana* (perhaps originally an intergrade between *L. m. mexicana* and the Blairi phase of *L. m. alterna*), the red extends down the side to near the belly and may actually continue onto the belly. The side and belly views below are of a very well-marked specimen.

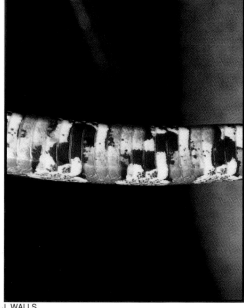

black spur. The red top of the head usually is present as a band but may cover the entire head. There are 196 to 212 ventral scales.

This phase has gained the hobbyist name of Variable Kingsnake because it often produces clutches that contain not only typically ringed specimens but blotched specimens, ones that look like *blairi* or *alterna*, and occasionally often are called the Durango Mountain Kingsnake by hobbyists and recognized as a distinct species or subspecies, *greeri*. Unfortunately they cannot be technically defined, and it has been known for over 30 years that snakes from this area are intergrades between *L. m. mexicana* and *L. m. alterna*. These snakes may have a mixture of red-centered black bands or blotches over the back and an

Durango Mountain Kingsnakes (*greeri*) come in many patterns. This specimen shows its Mexican Kingsnake ancestry anteriorly and its Gray-banded Kingsnake history posteriorly. Such mixed patterns are not uncommon in this intergrade phase.

some that are virtually indistinguishable from *triangulum* or *ruthveni*. How much of this variation is natural and how much is due to inbreeding and hybridization in captivity is unknown, but it is doubtful that *thayeri* exists as a distinct subspecies.

THE *GREERI* COLOR PHASE

Specimens of *L. mexicana mexicana* from southern Durango open nuchal blotch that may be large or small and connected or not with the rest of the head pattern, which may be simple (as in *alterna*) or complicated (as in *mexicana*). The ventral scale count is low, usually about 210 or fewer, like typical *mexicana*. This form cannot be defined scientifically and is a true mixture of almost any types of patterns that you want to put into it. I personally would recommend that

Head of an aberrant golden specimen with a greatly reduced pattern. The connection of the nape blotch with the head blotches indicates that this probably is *L. m. mexicana*, but it would be interesting to know its genetic history.

the name *greeri* not be used by hobbyists or dealers.

THE MYSTERIOUS *LEONIS* PHASE

In 1893, Günther described *Coronella leonis* from a single specimen almost 24 inches long from Nuevo Leon, Mexico. The specimen still is preserved in the British Museum and has been redescribed a couple of times, but apparently no specimen exactly matching it has ever been found again However, it seems evident that it is closely related to *L. mexicana mexicana*, and it has at times been synonymized with that form. Basically, it is a grayish

This hatchling looks a lot like a somewhat typical *alterna*, but it actually comes from a *greeri* line. However, it was sold as a *thayeri* variant, to make things even more confused. Keep in mind that names in these snakes may be almost meaningless.

Body view of the golden Mexican Kingsnake. I find this an attractive animal, but apparently it is not popular and has not been selectively bred.

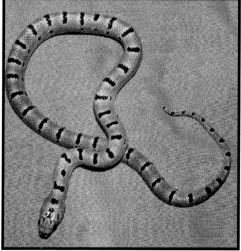

snake with 27 small, widely separated reddish blotches down the middle of the back. Each blotch is irregular in shape but generally oval and has a black outline. Some of the blotches are split down the middle and look like two round spots partially fused into a single blotch. There are only traces of small dark spots low on the sides, there is a short black bar back from the eye on the widened head, and the belly is pale with scattered black blotches. The nuchal blotch consists mostly of a black-outlined arch with long arms and traces of red within it. The nuchal blotch is widely separated from the large paired blackish spots that comprise most of the head pattern.

Strikingly similar patterns, though differing in details, occasionally appear in specimens intermediate between the *alterna* and *blairi* color phases, and I strongly suspect from the

A young Mexican Kingsnake, *L. m. mexicana*, showing signs of having some *alterna* genes in its ancestry. Such specimens often are sold as *greeri*.

Above and below right: Belly views of different *L. m. mexicana*. Details of color and pattern vary greatly in this subspecies.

reported locality that *leonis* represents a freaky intergrade between *L. mexicana mexicana* and *L. mexicana alterna*.

DISTINCTIONS

Usually, *L. m. mexicana* can be distinguished from *L. m. alterna* in its various phases by having fewer ventral scales (211 or less) than *alterna* (211 or more) and more body blotches (typically over 30 and often 40 versus usually fewer than 30 in *alterna*). Additionally, in *alterna* the nuchal blotch is widely separated from the head pattern, which typically is reduced to a few black spots or a short black Y or is absent; in *mexicana* the nuchal blotch extends further forward on the head and is connected by a black spur to the more complicated head pattern of paired black spots and bands. Both subspecies are so variable in pattern, however, that only a ventral scale count may adequately distinguish them.

CARE AND REPRODUCTION

Though the Mexican Kingsnake has its followers, most of the specimens seen for sale are sold

as the *thayeri* or *greeri* phases. Perhaps typical *mexicana* looks too much like a Corn Snake to appeal to more hobbyists. This subspecies seems not to differ from *alterna* in any aspects of its biology, captive care, and breeding. It is active at night and probably also during the early morning and evening and feeds on lizards (*Sceloporus, Cnemidophorus, Eumeces*) in nature. As usual for this group of kingsnakes, the young may be difficult to adapt to a mouse diet, and there are many losses before the snakes start feeding on their own.

The natural history of the Mexican Kingsnake is poorly reported, but it tends to inhabit rather dry mountain forests of oak and pine at the margins of drier, more desert-like areas. It is found at high elevations and often likes rocky canyons. Though there may be cacti in its habitat, this is not a desert-dwelling snake under normal circumstances.

It has been suggested that the color patterns of *mexicana* group kingsnakes have developed as mimics of various local venomous snakes. Some populations of *L. m. mexicana* are found in sympatry with the small rattlesnake *Crotalus triseriatus*, which bears a strikingly similar pattern of middorsal blotches though lacking the bright colors.

THE FUTURE

At the moment, the genetics of snakes sold as Mexican, Variable, and Durango Mountain Kingsnakes are so jumbled that any captive-breeding program involving specimens from the hobby is sure to lead to further confusion and a great mix of colors and patterns. There are no indications that the snakes are threatened in their natural habitat, so when Mexican regulations against export of specimens become relaxed, it should be possible to once again have purer stocks available. Hobbyists definitely are using too many names for the color patterns of this snake, and they are confusing the actually rather simple situation of variation and intergradation.

Two specimens of the Thayeri phase of the Mexican Kingsnake. When specimens are broadly banded with red like these, they often are called the "milk snake pattern."

PHOTO: I. FRANCAIS, COURTESY SHELTON.

GRAY-BANDED KINGSNAKES

In a period of just 30 years, the Gray-banded Kingsnake, *Lampropeltis mexicana alterna*, has gone from being one of the rarest and most poorly known Texas snakes to perhaps the most often-seen Texas kingsnake in the herpetocultural hobby. This is due to extensive collecting—many would say over-collecting—by literally dozens of collectors each year plus continuing captive-breeding by further dozens of hobbyists and commercial breeders. This snake also has one of the most complicated taxonomic histories of U.S. snakes, and even today there is great confusion as to exactly what to call individual specimens. This also is the only North American snake where hobbyists for some reason collect by locality rather than type.

HISTORY

When Brown described *Ophibolus alternus* in 1901, he had no idea what he was starting. The unique type specimen came from the Davis Mountains of Jeff Davis County in western Texas, today about an hour's drive south of Interstate 10, but in 1901 a hard, hot, dry expedition involving wagons and mules. Even today towns are few and far apart in this area and most are dead or dying. The land is overgrazed cattle ranches extending from horizon to horizon, usually with nothing but bare-looking sharp-topped mountains to block your

PHOTOS: M. & J. WALLS.

Above and below: Views of the Chisos Mountains in Big Bend National Park. This dry, hot habitat is the home of the typical phase of *L. mexicana alterna*.

The colorful Blair's phase of *L. m. alterna*, perhaps the most popular variety of Gray-banded Kingsnake.

PHOTO: R. MARKEL.

The cliffs of the Rio Grande at Big Bend. The narrow strand of beach is on the Mexican side of the river, the cliffs on the U.S. side.

Above and below: The habitat of *L. m. alterna "blairi"* in limestone crevices near Langtry, Texas. Here the snakes spend most of the year underground, coming out mostly during the summer rains.

view of forever. The geology of western Texas is complicated, but most of the region is part of the Chihuahuan Desert, with daytime temperatures of 110°F much of the year, dropping at night by 40 or 50 degrees from the daytime high. During the winter there may be much snow, with blizzards and high winds common. All told, this is one of the bleakest regions of the United States, but it certainly has a beauty of its own, including a diverse assortment of rare herps, birds, and mammals.

The second specimen of *L. alterna* (then still considered a full species) was not taken until 1938 in Casa Grande Park, Chisos Mountains, right in the "big bend" of the Rio Grande. The next specimen was reported by Smith in 1941 from Saltillo, Coahuila, Mexico. By 1957 specimens were beginning to appear more often, but there now was a complicating factor: a second species.

When Flury described *Lampropeltis blairi* from west of Dryden, Terrell County, Texas, he had good reason to think this specimen was truly distinct. The original *alterna* was defined by having alternate bands of narrow black scales, often broken into spots, and somewhat wider black bands split with red in the center of at least some bands. The nape spot was small or absent, and the overall-appearance of specimens from Jeff Davis and Brewster Counties, Texas, was of a gray snake encircled by black bands with a minimum of red in some of the bands. The new *blairi* came from considerably east of previous

GRAY-BANDED KINGSNAKES 57

In the Blair's phase the black rings are broadly split with red of various shades, usually with an orange tinge. This color pattern tends to occur in the eastern part of the range of the subspecies *alterna*. This specimen is from Juno Road in Seminole County, Texas, one of the more heavily collected localities. Not all specimens from Juno Road look like this one, however.

localities for *alterna*, part of the Pecos River basin in rolling limestone plains that, though dry and rocky, are far from being desert. The new *blairi* was a remarkably beautiful snake, grayish white with very broad, bright red saddles or blotches on the back, each outlined by a wide black line and a narrow white line. The nuchal or nape spot was very large and bright red like the body spots. When a second almost identical specimen was found by the Devils River in Val Verde County, even further to the east, the validity of the new species seemed confirmed.

The new rare species was from localities not far from San Antonio, and it was much easier for collectors to invade its habitat than that of the rougher *alterna* habitat, and collectors soon were taking more specimens from a variety of localities in Terrell, Val Verde, and even Edwards Counties. The species *blairi* appeared to be not uncommon, though hard to collect, in the Pecos-Devils Rivers area north of the Rio Grande. As you collected further to the west, approaching what is termed the Marathon Basin, a geological feature that helps define the Chihuahuan Desert and basically marks the difference between rolling limestone plains to the east and dry, rocky, desert to the west, oddly patterned snakes began to appear.

By 1965 herpetologists began to wonder about two variable kingsnakes occupying adjacent ranges and showing no differences in scale counts or structure. In 1970, Tanzer

A small female of the typical phase of *L. m. alterna* from south of Alpine, Texas. Notice that the dark bands alternate, a solid one followed by one that is broken into two or more spots.

PHOTO: K. H. SWITAK.

GRAY-BANDED KINGSNAKES

An interesting Gray-banded Kingsnake that shows a mixture of typical and *"blairi"* patterns. Both broad red saddles and narrow dark bands are present. The heavily marked head is favored by some breeders.

described the natural offspring from a pregnant *alterna* taken near Comstock, Val Verde County, far to the east in the range of *blairi*. Her offspring proved to include two rather typical *alterna* and three typical *blairi* color patterns, proving to everyone's mind that these two "species" are pattern polymorphs of a single variable form, *alterna*. Additionally, several workers had shown that *alterna* intergraded with *mexicana* in Durango and perhaps in Nuevo Leon, Mexico, so the name of the Texas snake had to stand as *Lampropeltis mexicana alterna*, used here.

In 1982, Garstka's famous paper attempted to re-elevate *alterna* to full species rank on the basis of small differences in hemipenis spines and eye color (silvery versus brown), but the opinion of people who know the snake well in life and nature is generally that he saw too few specimens and overemphasized minor characters. Though U.S. field guides often list *alterna* as a full species, there seems to be lots of evidence to show that *alterna* is not distinct at the specific level from *mexicana*.

DESCRIPTION

In structure, the Gray-banded Kingsnake is a rather small (24 to 32 inches, rarely to 4 feet and with a monstrous 5-footer recorded), slender snake with a long tail (about 15% of the total length) and a wide, flat head that is swollen posteriorly (the temporal region) and very distinct from the neck. The snout is short and wide, and the eyes are quite large and slightly bulging. Eye color often is a bright silvery gray, but also may be brownish. There is a broad black stripe back from the eye to the angle of the jaws in most specimens, though it may be reduced to a spot or absent. Important scale counts include: ventrals 211-230; subcaudals 56-67; dorsal scale rows 23-27 (usually 23-25); supralabials 7-8; infralabials 9-11. All the scale counts are higher than those for *L. mexicana mexicana*, but there is considerable evidence that scale counts in snakes are somewhat dependent on temperature and humidity during the incubation period, probably becoming genetically fixed over short periods of time. The ventral count should be sufficient to distinguish all typical *alterna* from typical *mexicana* outside the zone(s) of intergradation. The hemipenis spines are rather long for the group, 0.8 mm, and curved.

The color pattern is more difficult to describe, but basically there are two variable and broadly intergrading phases:

THE *ALTERNA* COLOR PHASE

Against a pale grayish (varying from whitish gray to distinctly bluish gray and sometimes very dark, almost blackish, gray) are narrow bands or rings of black that number between 15 and 39, with a most common number near 24. Typically one complete black band, which may or may not have a red center, alternates with a broken black band that may be reduced to a pair of spots or even a single spot at the middle of the back. Occasionally every black band has a bright red center, and also occasionally the entire back is covered with narrow, symmetrical black rings without any trace of red. In rare specimens the black rings are all or almost all broken into black spots, producing a speckled pattern. The scales of the gray background may be tipped with black to increase the speckled appearance. The nuchal blotch is short, weakly defined in many specimens, and may consist of a pair of black-bordered red spots or a larger reddish spot with an open gray center. Few specimens have more than 25 dorsal scale rows.

THE *BLAIRI* COLOR PHASE

The background color varies through the same intensities of gray, from almost white to bluish (sometimes bright blue) and brownish or almost black. The back is covered with broad bright red (sometimes orangish, sometimes brownish) saddles or blotches that are as many as 15 to 20 scale rows wide. These red saddles are outlined with black and then with a narrow white line. The nuchal blotch is long and red like a midbody blotch. Typically there are about 14 red saddles on the back. Dorsal scale rows are 27 more often than in

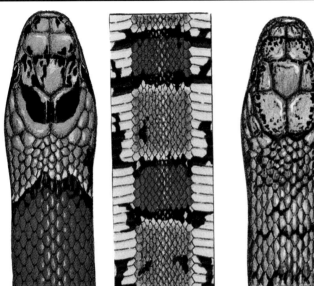
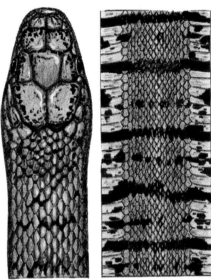

Diagrammatic representations of the heads and midbody patterns of the Blair's phase (left) and typical phase of *L. mexicana alterna*. Few specimens of the typical phase would match this exact pattern.

alterna, but this appears to be just a weak trend.

In both color phases the head pattern varies tremendously. Rarely it may be solid black (especially in captive-breds selected for the color), but more typically it has a scattering of black spots and lines that may form a short curve or Y and is separated by a wide distance from the nuchal blotch. Occasional specimens of both phases may have clean heads with few markings.

The belly color varies from being almost uniformly ringed to agree with the dorsal pattern to heavily blotched with black and red to almost clean white.

LOCALITY-MATCHED COLLECTING

The great variability of this snake has led to collectors and breeders restricting themselves to specimens from specific localities. I am not sure of the reason this trend has developed, but to each his own. Some hobbyists certainly believe that a particular pattern or coloration is restricted to one locality, but there seems to be no basis for this. One locality typically produces mostly one pattern, but there always are at least four or five variations on the basic pattern present if enough snakes are collected. It is not uncommon for snakes from any locality east of the Marathon Basin to vary between a common *blairi* type and a rarer *alterna* phase.

Commonly seen localities include:

Juno Road, Val Verde County,

basically Route 163 north of Seminole Canyon State Park.

Loma Alta, Val Verde County, the vicinity of a small town on Highway 277 just west of the Edwards County line.

Langtry, Val Verde County, the vicinity of the small town once the headquarters of Judge Roy Bean of movie and TV fame, located in beautiful rolling limestone hills and near the Pecos River Gorge, an incredible sight.

River Road, Presidio County, a road through public land at the edge of the Rio Grande west of the town of Lajitas. It is illegal to collect in Big Bend National Park, which occupies almost all the bend of the Rio Grande and the Chisos Mountains.

The first three localities produce beautiful *blairi* phase animals, while River Road produces *alterna* types that often are very speckled. Over 20 localities are available to the fanatical hobbyist who wants to spend the time to look for them, many being commercially bred in good numbers. Additionally, most color patterns are available as dark and light individuals.

COLLECTING

If you purchase a Texas hunting license, it still is possible to legally collect Gray-banded Kingsnakes on private property (with permission) and along small secondary roads that run through public property. Dozens of collectors, some very well organized, descend on western Texas during the spring and early summer, between April and June,

The speckled phase of *L. m. alterna* has the bands broken into often irregular spots over much of the body. This phase bears a strong resemblance to some local forms of the Rock Rattlesnake, *Crotalus lepidus*, found in western Texas.

especially when the rains come. At this time males leave their largely subterranean habitats in rock piles and limestone crevices to search for females, and even a few females can be found out and about. In the eastern part of the range the roads cut through limestone hills, leaving readily accessible road cuts that are searched between 10 P.M. and midnight using lights. Large fluorescent lights cannot legally be used along roadsides, but they are legal to use in canyons off the road if you have permission from ranch owners. Flashlights are legal at the moment, as are car headlights. Commonly a collector looks for a humid, moonless night on which to hunt, especially as a thunderstorm front approaches. He drives along the road until a remote section with many limestone cuts or rock piles is found and then either patrols the road with his car or walks along canyons searching with bright lights. In the process other herps will be found, especially rattlesnakes, sleeping swifts, and occasional other smaller snakes.

There have been several attempts to reduce collecting pressure on these snakes, but so far no truly restrictive laws have been enforced. You cannot collect in parks, especially Big Bend, or on private ranches without permission. Some areas enforce laws prohibiting hunting from a vehicle, so snake hunting from a car may get you in trouble if the laws are selectively enforced. Roads in this area are narrow and often winding, with many hills and poor visibility. Ranchers in pickup trucks know the roads and drive fast at night, which could lead to collisions with slow-moving "foreigners" looking for snakes. Be careful and be sure you have a license.

Today fewer snakes are taken in most areas than just five or six years ago, or so the rumors have it. These snakes normally do not eat mammals in nature, so attempts by reporters and law enforcement to place the blame on increases in rodent populations in the area on loss of Gray-bands are misplaced and erroneous. There is talk of increasing penalties for collecting from public roadsides and from cars, so be sure to stay alert to the most recent laws and their enforcement in the area you want to collect. Enjoy the scenery, which is beautiful throughout the area, and if you are in Langtry stop in at the Judge Roy Bean museum for a few minutes to get an idea of what the area was like less than a century ago.

CARE AND BREEDING

There is nothing much that I can say about keeping and breeding this subspecies that has not been mentioned in an earlier chapter. These are beautiful, docile snakes that are easy to care for and feed well as adults on a mouse diet. They adapt well to small cages with a variety of

PHOTO: M. & J. WALLS, COURTESY B. BARNES.

A yearling Gray-banded Kingsnake with a pattern intermediate between the Blair's phase and the typical phase (red saddles plus dark bands). Its parents are from Pecos County in the eastern part of the range.

subtrates, may live together well in small groups, and are tolerant of a variety of temperature and lighting conditions. To breed them you will have to cool them for about three months at 55°F and then reintroduce a female to a male in March or April. Eggs are laid about a month after mating (which usually occurs at night), and females often prefer to lay in a plastic box filled with damp peat moss. The eggs hatch in about 60 days (50 to 80) and produce delicate hatchlings that in nature would feed on sleeping lizards. In captivity they are hard to adapt to mice, and 50% mortality rates are not uncommon. Some young never adapt and must be forced-fed to keep them from starving. Other young will take pinkie mice from the tenth day, after their first shed, especially if individually confined in a small, dark cage. Sexual maturity may be reached in 18 to 24 months, but it is safest to not breed specimen less than 30 months of age. Adults may live 10 to 12 years, with some approaching 20 years.

DISTRIBUTION

Lampropeltis mexicana alterna is found in suitable habitat over much of the western part of Texas from the New Mexico line to the Mexican border, east to Edwards County. Its Mexican range is uncertain because the mountains of northern Mexico are hard to reach as there are few roads and facilities. It definitely occurs in Coahuila in suitable habitat and extends into Durango in the southwest, where it intergrades with *L. m. mexicana*. In the east its relationship to the *thayeri* phase of *mexicana* and to typical *mexicana* is uncertain, but it appears to share genes with these forms in southern Nuevo Leon. It is expected to occur in southern New Mexico above the Texas border, but for some reason it has never been found in the mountain ranges of that area. Specimens have been taken in Texas in sight of the New Mexico state line, and it is certain that the snakes must wander across at least occasionally even if there are no reproducing populations there.

This might be a good place to mention the theory that the color patterns of *alterna* have developed in mimicry of local venomous snakes, especially the equally variable Rock Rattlesnake, *Crotalus lepidus*, which often bears a striking resemblance to some individuals of both phases of the kingsnake. There also might be some mimicry of the local subspecies of the Copperhead, *Agkistrodon contortrix*. Of course, mimicry theories are hard to prove, and some scientists believe that mimicry of the type that would be found in the Gray-banded Kingsnake does not even exist. There is no doubt that some patterns of Gray-bands and local Rock Rattlesnakes share a remarkable similarity and the two snakes are found in similar habitats and often in very close association.

A yearling Gray-band from River Road in Presidio County. This pattern shows a great similarity to that of the holotype of *leonis* from Nuevo Leon, Mexico.

PHOTO: M. & J. WALLS, COURTESY B. BARNES.